The Propagation and Care of Plants

With Information on Various Methods and Tools for Propagating Plants

By

William T. Skilling

British Library Cataloguing-in-Publication Data
A catalogue record for this book is available from the
British Library

A Short History of Gardening

Gardening is the practice of growing and cultivating plants as part of horticulture more broadly. In most domestic gardens, there are two main sets of plants; 'ornamental plants', grown for their flowers, foliage or overall appearance – and 'useful plants' such as root vegetables, leaf vegetables, fruits and herbs, grown for consumption or other uses. For many people, gardening is an incredibly relaxing and rewarding pastime, ranging from caring for large fruit orchards to residential yards including lawns, foundation plantings or flora in simple containers. Gardening is separated from farming or forestry more broadly in that it tends to be much more labour-intensive; involving *active participation* in the growing of plants.

Home-gardening has an incredibly long history, rooted in the 'forest gardening' practices of prehistoric times. In the gradual process of families improving their immediate environment, useful tree and vine species were identified, protected and improved whilst undesirable species were eliminated. Eventually foreign species were also selected and incorporated into the 'gardens.' It was only after the emergence of the first civilisations that wealthy individuals began to create gardens for aesthetic purposes. Egyptian tomb paintings from around 1500 BC provide some of the earliest physical evidence of ornamental horticulture and landscape design; depicting lotus ponds surrounded by symmetrical rows of acacias and palms. A notable example of an ancient ornamental garden was the 'Hanging Gardens of Babylon' – one of the Seven Wonders of the Ancient World.

Ancient Rome had dozens of great gardens, and Roman estates tended to be laid out with hedges and vines and contained a wide variety of flowers – acanthus, cornflowers, crocus, cyclamen, hyacinth, iris, ivy, lavender, lilies, myrtle, narcissus, poppy, rosemary and violets as well as statues and sculptures. Flower beds were also popular in the courtyards of rich Romans. The Middle Ages represented a period of decline for gardens with aesthetic purposes however. After the fall of Rome gardening was done with the purpose of growing **medicinal herbs** and/or decorating church **altars**. It was mostly monasteries that carried on the tradition of garden design and horticultural techniques during the medieval period in Europe. By the late thirteenth century, rich Europeans began to grow gardens for leisure as well as for medicinal herbs and vegetables. They generally surrounded them with walls – hence, the 'walled garden.'

These gardens advanced by the sixteenth and seventeenth centuries into symmetrical, proportioned and balanced designs with a more classical appearance. Gardens in the renaissance were adorned with sculptures (in a nod to Roman heritage), topiary and fountains. These fountains often contained 'water jokes' – hidden cascades which suddenly soaked visitors. The most famous fountains of this kind were found in the Villa d'Este (1550-1572) at Tivoli near Rome. By the late seventeenth century, European gardeners had started planting new flowers such as tulips, marigolds and sunflowers.

These highly complex designs, largely created by the aristocracy slowly gave way to the individual gardener however – and this is where this book comes in! Cottage Gardens first emerged during the Elizabethan times, originally created by poorer workers to provide themselves with food and herbs, with flowers planted amongst them for decoration. Farm workers were generally provided with cottages set in a small garden—about an acre—where they could grow food, keep pigs, chickens and often bees; the latter necessitating the planting of decorative pollen flora. By Elizabethan times there was more prosperity, and thus more room to grow flowers. Most of the early cottage garden flowers would have had practical uses though —violets were spread on the floor (for their pleasant scent and keeping out vermin); calendulas and primroses were both attractive and used in cooking. Others, such as sweet william and hollyhocks were grown entirely for their beauty.

Here lies the roots of today's home-gardener; further influenced by the 'new style' in eighteenth century England which replaced the more formal, symmetrical 'Garden à la française'. Such gardens, close to works of art, were often inspired by paintings in the classical style of landscapes by Claude Lorraine and Nicolas Poussin. The work of Lancelot 'Capability' Brown, described as 'England's greatest gardener' was particularly influential. We hope that the reader is inspired by this book, and the long and varied history of gardening itself, to experiment with some home-gardening of their own. Enjoy.

THE PROPAGATION AND CARE OF PLANTS

Jock, when ye hae naething else to do, ye may be sticking in a tree; it will be growing, Jock, when ye're sleeping.

SIR WALTER SCOTT

Making a right beginning

THE old saying, " Well begun is half done," is a good motto for any one undertaking to plant a garden, a lawn, an orchard, or a field. Mistakes made in the beginning are hard to correct afterwards. Some common mistakes are failure to prepare the ground well before planting, using seeds and plants of poor quality or poor variety, and careless planting.

Five methods of multiplying plants

The principal methods of propagating (starting) plants are (1) planting seed, (2) planting cuttings (slips), (3) layering, (4) grafting and budding, and (5) planting bulbs. Whichever of these methods may be used, the most important thing is to start with seed or other material from the best possible parent plants. After that,

The selection of good parent stock

the most necessary matter is properly to care for the plants while they are young. But no amount of care will make possible the raising of good plants if one starts with poor stock.

Why seeds need to be tested

Propagation by seeds. In Chapter One we noted that each seed has a hard outer coat. Within this outer coat is the young plant or embryo, together with a supply of food on which the new plant lives until its roots and leaves are well started. The process by which the seed becomes a young plant — that is, by which it sprouts — is called " germination." Some seeds will not ger-

1

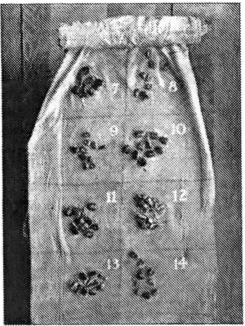

FIG. 43. The "rag doll" method of testing seed corn. The cloth, with the test grains in place, was rolled and tied, and it was kept thoroughly damp until the seeds germinated. The ten grains on each of the squares came from a separate ear that bore the number in the square. The ears numbered 8, 9, and 14 should not be planted.

minate at all; others germinate so feebly that the plant will either die or be weak and puny. Therefore seeds should be tested before they are planted (Fig. 43).

A convenient method of testing seeds is to count out *A simple* a certain number, as one hundred, spread them on one *test* end of a flannel cloth that has been wrung out of water so as to be damp but not saturated, and lay the other

2

end of the cloth over to cover the seeds. Place the folded cloth in a plate and turn over it another plate to prevent too rapid evaporation. Set it in a warm place and examine daily to remove all seeds that have sprouted. If as many as ninety seeds out of one hundred germinate, the seed may be considered fairly good. (Exp. 1.)

Storing seeds

If seeds are properly selected and cared for at harvest, and if they are properly stored till planting time, there will be little trouble in getting them to germinate.

Keeping fruit seeds moist

Most seeds should be kept dry until planting time, but those of apples, pears, peaches, and cherries, which are inclosed in a juicy fruit, should not be allowed to become dry after being removed from the fruit. If they remain dry long, the embryo dies. Such seeds should either be planted immediately or be kept damp until time to plant. They may be mixed with damp sand, put in a bag, and kept in a place so cool that they will not germinate until planting time; or they may be layered in sand. To layer seeds, cover the bottom of a box with damp sand; lay a cloth over this, and place a layer of seeds on the cloth; cover the seeds with another cloth, and on this place a new layer of damp sand. Proceed in this manner until the box is filled with alternate layers of seed and sand. Keep the box in a cool place.

Why seed corn should be kept dry

Seed corn should be kept in a cool, dry place. If it is allowed to absorb moisture or is not thoroughly dried after gathering, the moisture will freeze in the kernel and kill the germ. When the kernels are very dry, there is little danger that they will be damaged by freezing.

A good way to store seed corn is to remove the husks

Propagation and Care of Plants

FIG. 44. Stringing ears of seed corn.   *U. S. D. A.*

and string the ears, hanging them from rafters or from *Ways of* a wire; or they may be strung as shown in Figure 44 *storing* or stuck on spikes as in Figure 45. Another good way *corn* is to turn the husks back, braid them together in bunches, and then suspend the corn. The ears may also be laid in racks, but there they are likely to be attacked by rats and mice. Under no circumstances should the ears be allowed to touch each other. (Exp. 2.)

Whether seeds are planted in the open field, the *The seed* garden, or the greenhouse, the method of preparing the *bed* seed bed is the same in principle. The chief difference lies in the degree of care that it is possible to take. In all cases the seed bed should be (1) deep, (2) well drained, *Eight* (3) mellow, (4) well packed below the surface, but (5) *conditions* fairly loose on top, (6) free from clods, (7) as level as *desired*

4

Prepara-
tion of seed
bed

Garden
soil

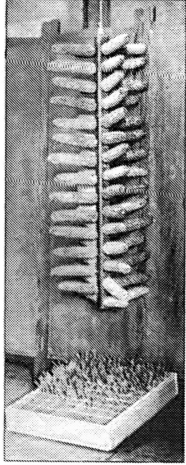

Fig. 45. Seed corn stuck on spikes.
Notice the seed-test box below.

possible, and (8) composed of soil well supplied with plant food.

The process used in getting the seed bed into proper condition will depend upon its size. Deep plowing in the field, deep spading in the garden, and filling a box with fresh soil in the greenhouse will all accomplish the same purpose,—they will mellow the soil and admit air, giving the roots a chance to penetrate to a good depth. Clods should be broken up immediately, except after fall plowing, which is left rough over winter.

Proper richness of soil cannot always be had in large fields without great expense; but in a garden, where so much and such expensive seed is used and where so much labor is expended, it is false economy to save on fertilizer.

Propagation and Care of Plants

A soil suitable for seed boxes ought to have sufficient sand and humus to keep it from forming a crust. One-third ordinary soil, one-third sand, and one-third leaf mold makes a good mixture. *The right soil for seed boxes*

Packing the soil, as we have seen, is accomplished in the field by means of a roller or a subsurface packer. In the greenhouse a flat wooden block called a " float " is used to firm the earth in the seed boxes. The surface in both cases is afterwards made loose, — in the field by harrowing and in the greenhouse by sifting a layer of fine soil on top of the packed soil in the seed boxes. In the garden a rake is used in making the mulch. *Packing the soil* *Tools used*

A rule often given for greenhouse planting is this, — that a seed should be covered with a layer of soil equal to its own thickness. Thus a coconut might be placed about seven inches underground, but such fine seeds as those of the begonia should have as little covering as possible. They are sometimes covered only with a damp cloth, which is removed when the seeds are sprouted. *Depth of planting; a greenhouse rule*

A small seed must be near the surface because the new plant cannot begin to draw nourishment from the air until it comes up into the light. If it has far to come, it may die of starvation before it reaches the surface. A large seed, having a greater store of nourishment, may send up a shoot through several inches of soil before the young plant needs to depend upon food of its own making. *Why small seeds must be near the surface*

In the field and garden, where the surface of the ground is very dry and moisture from below is relied upon to cause germination, the seeds must be put deep enough *Rules for field planting*

for the capillary water to·reach them. (Exp. 3.) In
cool, wet weather, however, seeds need the surface warmth
which the sunshine supplies, and they will rot if too
deeply covered. The looser and drier the soil, the
deeper the planting should be. The usual depths for
planting are: clover and grass seeds ½ to 1½ inches;
wheat, oats, etc., 1¼ to 3 inches; beans, 1 inch; peas
and corn, 2 to 4 inches. The smaller depths above
given are for heavy, moist soils; the greater depths are
for light, dry soils.

*Proper
spacing;
the danger
of planting
too thickly*
Planting too thickly often accounts for poor results.
If very close together, plants interfere with each other's
growth and become weak and spindling, just as they do
if they are crowded by weeds. If plants come up too
thickly they should be thinned at once, before they
begin to crowd each other. In doing this, keep in mind
the size that the plant will have when full grown. Large
seeds are seldom planted too thickly; but great care has
to be used in planting small seed like that of turnips.
One good way to handle such seed is to mix it with fine,
dry sand or dust before planting. Another method
often employed with cabbage, onion, and lettuce seed
is to sow it thickly in boxes of earth and transplant
before the plants are big enough to crowd each other.
In China, where labor is cheap and land is scarce, this
method is employed in raising rice.

Propagation by cuttings. A "cutting" or "slip"
is a piece of a small branch or twig planted to take root.
Even a leaf of some plants will take root and grow
(Fig. 46). Most of our house plants, and roses, and such
fruit plants as currants and grapes, are propagated by

cuttings. One advantage in propagating by cuttings is that growth is generally more rapid than from the seed. Another advantage in their use is that the plants produced will be like the parent plants, whereas seedlings are often very different, especially in fruits and flowers.

Most trees and shrubs and vines — plants of hard and woody fiber — grow best from cuttings made of wood that has

W. T. Skilling

FIG. 46. A leaf cutting of begonia.

had a year's growth. These are known as " hardwood cuttings " (Fig. 47). On the other hand, herbs such as verbenas, fuchsias, and petunias — soft and pulpy plants — grow best if the cuttings are made from the soft tips of growing branches. Such slips are called " softwood cuttings " (Fig. 48).

A hardwood cutting should be long enough to include at least two nodes (the place from which a bud grows), — one to produce roots, and one for branches to grow from. The lower cut should be made just below a node, and the upper cut, if there is to be one, between two nodes. Softwood cuttings should be very short, including only one bud.

Nature-Study Agriculture

Cuttings must be made with a very sharp knife. A clean cut heals over more easily than does a torn and

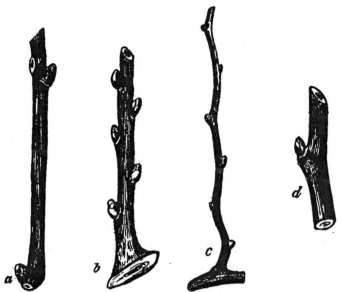

FIG. 47.   Hardwood cuttings: *a*, simple cutting; *b*, heel cutting; *c*, mallet cutting; *d*, single-eye cutting.

slivered end;   and unless the wound quickly becomes " callused " over with new growth, bacteria will soon enter and cause decay.   All leaves that would be covered and a part of those that would be above ground should be removed.   Covered leaves might start decay, and too many above ground would draw moisture from the cutting and dry it out too much.   (Exp. 4.)

Planting cuttings; why sand is used   In the planting of cuttings it is necessary to keep in mind the need for air.   It is especially desirable for the production of roots that plenty of air be present, and for

9

Propagation and Care of Plants

this reason cuttings should be started in coarse, damp sand (Fig. 49). No other nourishment than air and moisture is necessary for getting the roots started, and as sand admits air more freely than other soil does, it is better to use either that or a very sandy loam. Slips *The effect* are often prevented from forming roots by being kept *of too much* *water* too wet. Even sand, if water-soaked, will not admit air; and there must be proper drainage or the cuttings will rot. (Exp. 5.)

About three fourths of the length of a cutting should *The* be placed below the surface. The purpose in planting *purpose* *in planting* so deeply is to prevent drying out. A light-colored *deeply* cloth screen laid loosely over a bed of slips helps to keep the tops moist. A glass jar is sometimes set over the plant to prevent loss of moisture.

Some plants, such as the geranium and the Wandering Jew, grow so easily from cuttings that little care is necessary. But with most plants roots cannot be so easily induced to start, and if we would succeed well we should follow the best methods.

Cuttings are most easily removed from the sand bed as soon as they have callused over at the cut, but before roots have actually appeared. (After the cut has healed by being callused over, roots are quite certain

When to *transplant*

U. S. D. A.

FIG. 48. A softwood cutting. (Slip of coleus.)

FIG. 49. Starting cuttings of geranium. Geraniums are among the easiest plants to propagate in this way.

to develop.) Cuttings from roses often die if they are disturbed when the roots are young. So it is better to plant rose slips in the open ground with a shovelful of sand in each hole, leaving them for a year before transplanting. They will then be thoroughly rooted.

The advantage in layering

Layering. As a cutting must depend upon the nourishment within itself until it can produce roots, it must be protected very carefully from influences that would cause it either to dry out or rot. But if a twig can be left joined with the parent plant while it is forming roots, there is little danger of failure (Fig. 50). This can be done by bending the branch or vine down to the ground, covering a part of it in the moist earth, and letting it remain there until roots develop and a new plant is formed. This process is called " layering." Bushes and vines, as black raspberries and grapes, may be started in this way. (Exp. 6.)

11

Propagation and Care of Plants

In layering, as in preparing cuttings, the leaves and buds which would go underground should be removed. The covered portion of the stem must be fastened securely to prevent it from being moved by the wind. A shovelful of sand or leaf mold mixed with the earth at the point where the branch is buried will make it root more easily.

A notch is sometimes cut at the point where it is desired to force out roots, as they develop more easily at a cut. A better method is to split the branch (Fig. 50, *a*) and put dirt into the opening. In ring layering (Fig. 50, *b*) advantage is taken of the fact that sap cannot run back toward the roots if a ring of bark is removed; that is, if the branch is girdled. Nourishment from the roots can flow past the girdle, for it travels within the stem. But the returning sap flows

Methods of encouraging roots to start

Ring layering

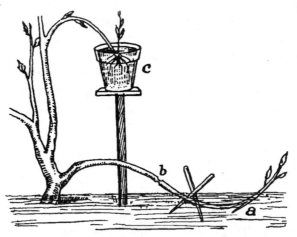

Fig. 50. Layering: *a*, stem split to start roots; *b*, ring of bark removed; *c*, "tip layering" in pot. Notice the crossed stakes which hold down the stem that has been laid underground.

12

in the inner layer of bark and so is checked at the girdle, concentrating nourishment at the point where roots are to be produced.

Layering in a box

If branches are too high to be layered in the ground, a box of earth in which to start the new plant may be supported as shown in Figure 50, *c*. Such a box or pot may also be used for branches at the ground; then, when roots form, the plant is already potted and is easier to remove.

Offsets; runners; rhizomes; suckers; division

Various methods of propagating. Bulbous plants such as tulips and hyacinths form new shoots on their roots which can be separated and replanted. These are called." offsets."

Plants like the strawberry have creeping branches which send roots down into the earth at intervals, thus starting new plants. Such branches are called " runners" or " stolons " (Fig. 51).

Rhizomes are underground runners. They are not roots, but are rootlike stems which send up shoots here and there, as in the iris.

Suckers are young plants that grow from the roots of the parent. They may be seen coming up around an elm tree. Some blackberries are propagated by transplanting suckers. Suckers from grafted trees are not good to transplant, for they will be like the seedling upon which the grafting was done.

Division is a good means of propagating some plants, particularly ferns. A plant is divided through the roots or rhizomes, and the parts are set out as separate plants.

Advantages in grafting

Grafting and budding. In grafting, a branch several inches long from one tree is inserted into another. If

Propagation and Care of Plants

FIG. 51. A strawberry plant that was set out in the spring at Osage, Iowa, photographed in the middle of July. Note the runners in bloom.

a bud instead of a small branch is grafted into another tree or shrub, the process is called " budding " (Fig. 52). In either case the fruit produced on the branches growing out of the graft will be exactly the same as the fruit of the tree from which the graft was taken. In fact, the " scion," the new growth from the point at which the graft was made, is but a continuation of the original tree from which the scion was taken, not a new tree as is one raised from a seed. Seedling fruit may or may not be good, but we can know the quality of grafted fruit beforehand. This was noted as an advantage in the use of cuttings. But in grafting, a double advantage can be secured, since the root which is to support the tree may also be selected. Since this is so, nurserymen try to graft the best fruit-bearing branches upon *How scion and stock are selected*

14

W. T. Skilling

FIG. 52. White roses budded on a red rose bush. All characters of the budded roses — as shape, size, and color — are the same as they would be if the roses were growing on their own bush.

varieties that have the hardiest roots. It must be remembered, however, that only closely related trees can be grafted — as peaches on plums or apricots, but not peaches on apples. The plant into which the scion is grafted is called the " stock."

Putting the growing layers in contact The process of grafting is not very difficult, but budding is so much easier that it is more commonly practiced than grafting. (Exp. 7.)

In either grafting or budding, the most important thing is to see that the growing layer of the bud or graft is placed in contact with the growing layer of the stock. The growing portion of a tree is the layer that lies just under the bark. This is called the " cambium

Propagation and Care of Plants

layer." The method by which cambium layers of stock and scion are brought together is made clear in Figures 53, 54, and 55.

The " tongue " or " whip " graft is very commonly *The tongue graft* used. Scion and stock must be of about equal size, the diameter of a lead pencil being about right, and cuts should be made as shown in Figure 53. In uniting the scion and stock, care must be taken to have the growing layers of the two stems in contact, at least on one side.

In the " saddle graft " the stock is cut wedge-shaped ; *The saddle graft* the scion is split, placed over the stock, and firmly pressed down. A saddle graft is easier to make than is a tongue graft, but it has rather less cut surface where growth may take place. (Exp. 8.)

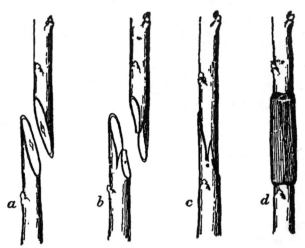

FIG. 53. The four steps in whip grafting: *a*, the first cuts; *b*, the cut surfaces extended to increase contact of cambium layers; *c*, scion and stock joined; *d*, the graft bound up.

Grafting is usually done when the stock is small, one year old or less, but sometimes an old tree is " worked

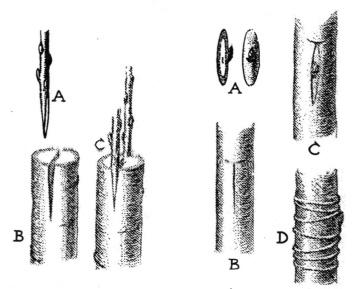

Figs. 54 and 55. Cleft grafting (left) and budding (right). In each figure, *A* is the scion and *B* the stock; *C* shows scion and stock joined.

over " or " top grafted," in order to make it bear more desirable fruit. In this operation " cleft grafting " is used. All the limbs are sawed off, the ends are split or sawed open, and a graft is wedged into the end of each limb. Sometimes two grafts are inserted (Fig. 54). Unless the graft is firmly wedged in place, it must be tied with raffia or soft cord. (Raffia is a fiber that comes from a kind of palm. It may be purchased at seed stores.) The binding material must be removed as soon as growth begins. When the limbs are taken

Propagation and Care of Plants

off, the tree should be protected from sun scald by a coat of whitewash. Often a few limbs are left until the next year to shade the tree.

In all grafts, wax is spread over the place where the *Grafting* stems have been joined. This is done to keep out air *wax* and moisture. Where slender stems are grafted, they are generally bound together before the wax is applied. Good grafting wax can be made by melting together one pound of tallow, two pounds of rosin, and a pound of beeswax.

The most common method of budding is illustrated *Budding* in Figures 55, 56, 57, and 58. It is known as " shield " or " T " budding. For this work buds should be selected *Selection* from the middle part of the branch. Those at the tip *and removal* *of bud* are too young, and those far back are too old. They should be dormant (resting); that is, they should not have begun to unfold. A well-developed bud is chosen, and the edge of a sharp knife is placed crosswise half an inch above it. A cut is made downward through the bark till the knife comes out half an inch below the bud. With the point of the knife some of the wood is picked

from the back of the bud. If there is a leaf at the bud, it should be cut off so as to leave the petiole for a handle. A T-shaped cut is made in the stock; the bark on each side of the vertical cut is rolled back, and the bud is inserted (Fig. 57).

A wet strip of raffia is

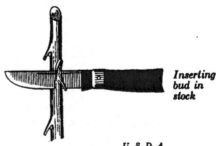

Inserting bud in stock

U. S. D. A.

Fig. 56. Cutting a bud.

W. T. Skilling

Fig. 57. Budding seedling fruit trees; inserting a bud, and wrapping with raffia.

Binding it   used to bind the bud in place. Care should be taken to cover the cut entirely with the raffia so as to exclude air and prevent drying. Sometimes a little grafting wax is smeared over the cut, but this is usually not necessary. The raffia should be removed as soon as the bud shows signs of growth, which will usually be within a month or six weeks.

Any sprouts that may grow on the stock after budding should be removed. The bud will then grow more rapidly, as it will receive all the nourishment that the root system produces (Fig. 58).

The time for budding and grafting   Grafting is usually done in the spring when the sap begins to flow. June and July are considered good months for budding; but this may be done successfully at any time when there is sap enough to make the bark loosen easily. (Exp. 9.)

19

Propagation and Care of Plants

Caring for trees. A seedling fruit tree may be budded when it is about a year old. It is then left in the nursery until the bud has had a year to grow, when the tree may be higher than a man's head and have many branches. Upon removal to the orchard, it is at once pruned

FIG. 58. When a bud has grown fast, the stock is generally cut away, just above the bud.

FIG. 59. Peach trees pruned, ready to plant.

back (Figs. 59 and 60). The central stem is cut off (headed) about thirty inches above the ground; only four or five of the strongest limbs are left, and even these are usually much shortened. Every cut should be made just beyond a bud, so that no stubs will be left to

U. S. D. A.

Fig. 60. Peach tree six months after planting.

die. The manner of pruning a peach tree until it is ready to bear is shown in Figures 61, 62, and 63.

After a fruit tree is given the desired shape by the first few years of pruning, it needs little attention other than to be kept from growing so many branches as to crowd the fruit or keep out the sunlight (Fig. 63). Any dead branches should be trimmed out.

Protecting orchard trees from frost

There are a few weeks in early spring when fruit buds are liable to be injured by frost. Many orchardists have stoves ready to light on frosty nights. The temperature of an orchard may be raised several degrees by these outdoor fires (Fig. 65).

An orchard on low ground is in much greater danger of being injured by frost than is one on high ground (Fig. 66). Cold, like water, flows to the bottom of a valley. A sharp narrowing of the walls of a valley, or a grove of tall trees within it, will check the flow of air as water is checked by a dam. Such a place, where cold air cannot drain away, is badly chosen for an orchard. (Exp. 10.)

If a tree is injured by animals or disease or by being

21

Propagation and Care of Plants

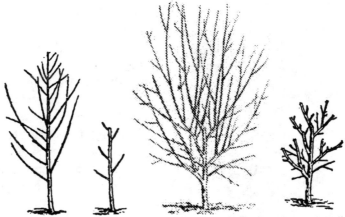

FIGS. 61 and 62.   First year in orchard: Fig. 61 (left), branched yearling, and same tree cut back at planting; Fig. 62 (right), first summer's growth in the orchard, and first winter pruning (December).   Compare Figures 59 and 60.

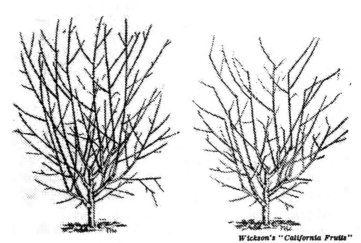

FIG. 63.   Second year in orchard: second summer's growth (September); second winter pruning (December).

Nature-Study Agriculture

FIG. 64. Pruning tools: curved blade saw, hedge shears, and long- and short-handled lopping shears.

Tree surgery broken by the wind, it may often be repaired so that it will recover from the injury. Figure 67 shows methods of " tree surgery," which may add many years to the life of a valuable tree.

Experiments and Observations

1. Test one hundred of each of several different kinds of seeds.

2. Before the first heavy frost, select several of the best ears of corn you can find and store them in one of the ways described in this chapter.

3. To test the proper depth of planting, plant in the garden a few seeds of each of several different crop plants: barely cover a few seeds of each variety; cover a few of each half an inch deep, a few of each an inch deep, then others two, three, four, five, and six inches.

4. See how many different kinds of plants you can make grow from cuttings, following the directions given in this chapter.

Propagation and Care of Plants

FIG. 65.   Oil heaters in use in an orchard.

FIG. 66.   One morning, at daybreak, the Fahrenheit thermometer registered 39 degrees above zero on top of this bridge, and there was, of course, no frost at that level; but in the valley below the thermometer registered 29 degrees, and there was a heavy frost.

24

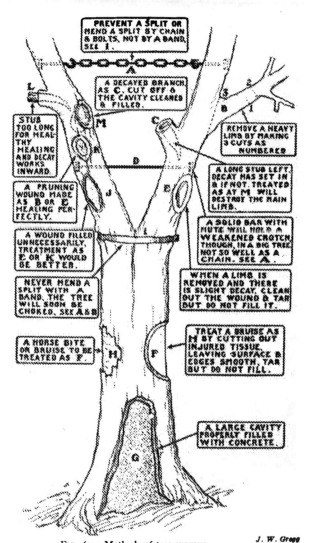

Fig. 67. Methods of tree surgery.

J. W. Gregg

Propagation and Care of Plants

5. Prove the need of underdrainage by planting seeds in two cans, only one of which has holes punched in the bottom. Keep both thoroughly watered.

6. Try to make vines take root by layering them.

7. Following the directions that have been given, see if you can successfully bud and graft rose bushes or young trees.

8. Start a little nursery at school or at home by planting peach pits in spring. The next spring, bud the seedlings.

9. Make a list of all the ways by which plants are multiplied, as mentioned in this chapter. See if you can find some example of each in your neighborhood.

References

"School Exercises in Plant Production." Farmers' Bulletin 408.
"Pruning." Farmers' Bulletin 181.